BEE AWARE
Count the Numbers

Written and Illustrated by
B.K MACKLE

First published in Far North Queensland, 2025 by Bowerbird Publishing

© 2025 B Kay Mackle. All rights reserved.

No part of this publication may be reproduced, stored in a retrieval system, or transmitted in any form or by any means - electronic, mechanical, photocopying, recording, or otherwise - without the prior written permission of the publisher, except in the case of brief quotations used in reviews or scholarly analysis.

This work is protected under Australian Copyright Law and international treaties.

Use of this content for the purpose of training or developing artificial intelligence (AI) systems, including large language models, machine learning, or data mining, is strictly prohibited without explicit, prior written consent from the copyright holder.

ISBN 978 1 7640739 5 0 (print)
ISBN 978 1 7640739 6 7 (ebook)

Bee Aware
Count the Numbers
Written & Illustrated by B Kay Mackle

Printed by MacPrint Mackay

Distributed by Bowerbird Publishing
Available in National Library of Australia

Bowerbird Publishing
Julatten, Queensland, Australia
www.crystalleonardi.com

As I listen,
I can hear buzzing in my ear.
What can it be?
A honey bee!

As I look,
I see,

1

ONE HONEY BEE

As I look closer,
I can see,

**1 - 2 - 3
ONE - TWO - THREE**

I can count - I can see HONEY BEES.
Makes me happy as can be.
I can see,

THREE HONEY BEES

As I listen,
I can hear a whisper in my ear.
What can it be?
Leaves on a tree!

As I look,
I see,
Flowers on trees.
As I look closer,
I can see,

1 - 2 - 3 - 4 - 5
ONE - TWO - THREE - FOUR - FIVE

I can count,
I can see,
TREES - HIGH-FIVE YIPEE!
Makes me happy as can be.
I can see,

FIVE GUM TREES

All swaying in the breeze.

As I listen,
I can hear,
Sweet songs in my ear.
What can that be?
A Bird on a tree.

As I look,
I see,

1
ONE FAIRY WREN

As I look closer,
I can see,

1 - 2 - 3 - 4 - 5 - 6 - 7 - 8 - 9 - 10
ONE - TWO - THREE - FOUR - FIVE
SIX - SEVEN - EIGHT - NINE - TEN

I can count,
I can see,
FAIRY WRENS
Makes me happy as can be.
I can see,

TEN FAIRY WRENS

I can see butterflies,
Too many to count.

**1 ONE - 2 TWO - 3 THREE - 4 FOUR
5 FIVE - 6 SIX - 7 SEVEN - 8 EIGHT
9 NINE - 10 TEN - 11 ELEVEN
12 TWELVE - 13 THIRTEEN
14 FOURTEEN - 15 FIFTEEN
16 SIXTEEN - 17 SEVENTEEN
18 EIGHTEEN - 19 NINETEEN
20 TWENTY**
butterflies fly free in the air,
being as quiet as one butterfly can be.

Twenty I see.

BEES, BIRDS, BUTTERFLIES,
FLOWERING GUM TREES.
I can see all these HEALTHY SPECIES.

Makes me happy as can be.
I can see them swaying FREELY in the AIR.
I can see them WORKING TOGETHER and
LIVING HAPPILY.
Without a care.

Together the bees make HONEY,
for us to SHARE.

BE AWARE.
The PLANET
Needs the BIRDS and the BEES.

THE FLOWERS and the TREES and
BUTTERFLIES.
Too many to describe.

WHY?

We rely on each other to survive.

LET US WORK TOGETHER,
CONTINUE TO LIVE IN HARMONY
with ALL that there is on EARTH.

ABOUT THE AUTHOR

Kay Mackle believes she is a messenger for the children and in her own way she hopes to assist the earth's environment and its people to regain balance in health and energy. Kay has been working with energy since 2011 to gain and raise awareness of her purpose in this lifetime.

Kay lives in Mackay Queensland Australia. She is available to assist others heal with Intuitive energy healing. She is a certified Reiki practitioner. She acknowledges her guides that have shown and educated her in the knowing that there are no coincidences. Everything happens for a reason. Health is Wealth. Healing takes time. Time is infinite. Love is an intense feeling of affection that is felt from within the heart. Everything happens in divine order and in divine time. Her spirit and her soul have been challenged. She has risen to great heights by allowing herself to grow by finding strength and courage to overcome many obstacles in her life.

Kay is inspired by and finds happiness and joy in earth's garden.

Kay Mackle
Author and Illustrator

FROM THE PUBLISHER

Bee Aware is a gentle reminder to slow down, observe, and reconnect with nature.

Author and illustrator Kay Mackle delivers a heartfelt tribute to the wonders of the natural world. More than just a picture book, this title is a sensory invitation - encouraging children, parents, grandparents, and educators to rediscover the joy of reading, counting, and observing together.

This book is particularly valuable for families and classrooms looking to introduce children to basic ecological awareness through interactive learning. Bee Aware is a delightful and meaningful addition to any school library.

Congratulations Kay, on yet another beautifully written and illustrated book.

Crystal Leonardi
Bowerbird Publishing
www.crystalleonardi.com

OTHER TILES BY B.KAY MACKLE

Titles of relevant interest for children of all ages:
THE POWER OF THREE (Darkness to Light)
THE PURPOSE OF THREE (In A Garden of Choice)
TIME TO HELP BUNDY BLUE
MASTER FOREST WIZARD

For older readers:
THE BROKEN BRANCH GROWS
THE GREY WOLF

www.ingramcontent.com/pod-product-compliance
Lightning Source LLC
Chambersburg PA
CBHW081412070526
44583CB00020B/2782